Health
Grade Three

Table of Conten

The school curriculum designates the topics taught to students, with the most stress placed on the core subjects of language arts and math. Health is the subject usually taught in the spare days between units in order to meet standards. However, health teaches students valuable life skills that they need to learn to develop healthy living habits. These skills will help students make responsible choices that affect their daily lives.

In this series, *Health,* information was compiled to supplement the required standards for health in each grade level. The activities were selected to complement the core subjects. So, instead of fitting health into the curriculum in one or two specific units, it can be introduced in any subject throughout the year. (The Curriculum Correlation chart on page 2 will help you determine which activities to incorporate into your core subjects.) The activities are fun and challenging, thereby avoiding the stigma of "boring" health worksheets. Many activities are open-ended so that students will have to think about, evaluate, and apply healthy habits to their daily lives. Moreover, introducing the activities into the core classes may reinforce positive habits.

Health is divided into three units. Unit 1 provides information on nutrition. It includes the food pyramid, food nutrients, and ways food affects teeth. Unit 2 explores safe health practices when outdoors. Here, students learn about such topics as sunburns, water safety, harmful insects and plants, and basic first aid. Unit 3 introduces the importance of life-long health habits, such as exercise, dental and physical hygiene, checkups, and elimination of germs. As students learn about health more often, it serves as a reminder of healthy habits. They will more likely remember the information, and thereby, apply it to their daily lives—the goal of any health program.

Health
Grade Three
Curriculum Correlation

Activity	Math	Language Arts	Social Studies	Science	Physical Education
Unit 1: Nutrition					
The Food Pyramid	collect data geometry	use pictures		human body	
A Balanced Diet	collect data time	use pictures		human body	
Nutrients for Your Body	use charts	vocabulary comprehension		human body	
Foods and Their Nutrients	categorize	use pictures			
Nutrition Scavenger Hunt	categorize	read for detail	community		
Reading Food Package Labels	collect data compare data percentages	read for detail			
Fat Test	use charts			experiments	
Snacking Test	use charts				
Healthy Teeth	collect data time			human body	
Calories	measurement compare data				
Energy		use pictures			exercise
Trail Mix	measurement				exercise
Unit 2: Safe Outside!					
Sunburns		comprehension		energy	
Insect Pests		use pictures		animals	
Insect Pest Safety		use pictures		animals	
Mosquitoes		comprehension		animals	
Healthful Hiking		comprehension	geography		exercise
Poison Plants		comprehension use pictures		plants	
Plants—Harmful and Helpful		comprehension	geography	plants human body	
Thunderstorm Safety				energy	
Water Safety Rules		context clues			exercise
Bicycle Safety Rules		using pictures sentences			exercise
Bike Reflectors				experiments energy	
First Aid		comprehension vocabulary		human body	
Unit 3: Healthy Habits					
Exercise and Sleep				human body	exercise
Warm Up and Cool Down	use charts			human body	exercise
Exercise and the Heart	use charts measurement	comprehension		space human body	exercise
Good Hygiene		sentences use pictures		human body	
Dental Hygiene	use charts			human body	
A Checkup		summarizing	community	human body	
Germs That Cause Diseases		vocabulary		human body	
Using Bacteria to Stay Healthy	use charts measurement			human body	
Changing Your Health Habits	use charts				

Health Assessment

· ·

✎**Darken the circle beside the answer that correctly completes each statement.**

1. A _____ diet gives you the right amount of food from the five basic food groups.
- Ⓐ fat
- Ⓑ working
- Ⓒ balanced
- Ⓓ fruit

2. Beans and nuts are from the _____ food group.
- Ⓐ meat
- Ⓑ bread/cereal
- Ⓒ milk
- Ⓓ vegetable

3. A _____ is a food nutrient that gives you energy to think and play.
- Ⓐ protein
- Ⓑ carbohydrate
- Ⓒ vitamin
- Ⓓ fat

4. Some unhealthy snacks are _____.
- Ⓐ fresh fruits and vegetables
- Ⓑ potato chips, candy, and soda
- Ⓒ bread and peanut butter
- Ⓓ cheese and crackers

5. To keep teeth and gums healthy, you should _____.
- Ⓐ eat sweets
- Ⓑ never brush
- Ⓒ floss
- Ⓓ chew on ice

6. Your heart beats faster when you _____.
- Ⓐ run
- Ⓑ sleep
- Ⓒ read a book
- Ⓓ watch television

Go on to the next page.

Health Assessment, p. 2

7. _____ can prevent you
   from getting a sunburn.
   Ⓐ Sunglasses
   Ⓑ Sunscreen
   Ⓒ Cold compresses
   Ⓓ Swimming

8. You can keep insects away
   by _____.
   Ⓐ wearing perfume
   Ⓑ walking barefoot
   Ⓒ wearing light-
      colored clothing
   Ⓓ eating picnic food

9. The oil from _____ can
   cause a skin rash.
   Ⓐ poison ivy
   Ⓑ oak trees
   Ⓒ foxglove plants
   Ⓓ evergreen trees

10. It is very important to wear
    _____ when riding
    a bicycle.
    Ⓐ knee pads
    Ⓑ a backpack
    Ⓒ dark clothing
    Ⓓ a helmet

11. Exercise makes the _____
    grow big and strong.
    Ⓐ feet
    Ⓑ hair
    Ⓒ arms
    Ⓓ muscles

12. To stay healthy, _____.
    Ⓐ do not wash with soap
       or shampoo
    Ⓑ do not wash your hair
    Ⓒ wash all parts of your body
    Ⓓ wash only your hands

The Food Pyramid

You are waiting for the lunch bell to ring. Why are you so hungry? Well, you had a busy day. You walked to school, hurrying so that you wouldn't be late. Those exercises in gym class were fun, but tiring. You also worked hard in your other classes today. You used up a lot of energy. No wonder you are looking forward to lunch.

Look at the food pyramid. It shows the five basic food groups. Foods in the same food group provide about the same nutrients for your body. Did you know that nutrients are substances in food that you need so that you can grow and have energy?

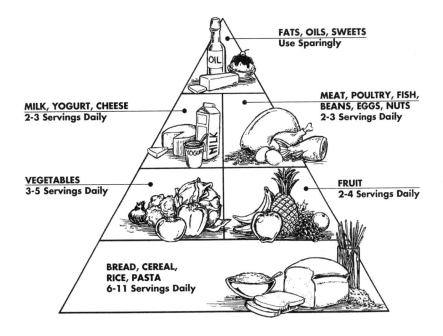

✎What did you eat for your last meal? Write down each food. Then tell which food group it is from.

_____

_____

_____

_____

A Balanced Diet

Your diet is the combination of foods you eat each day. A *balanced diet* gives you the right amount of food each day from the five basic food groups. After you have made your food choices from the five basic food groups, you might make additional choices. You might decide to add items that include fats and oils from the top of the pyramid. Foods that have fats and oils include cakes, cookies, and french fries. Eating too much of them can cause you to gain extra weight and may be harmful to your teeth.

Scientists and doctors continue to do research about food and share their findings. More than ever before, people are concerned about what they eat. They are finding out that a balanced diet can be an important part of good health.

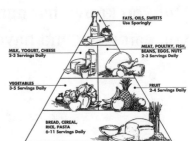

✎Keep a record of what you eat for one day.

Breakfast	Lunch	Dinner

✎Did you eat a balanced diet? Tell how you can eat a more healthy diet.

_____

_____

_____

_____

Nutrients for Your Body

Food contains *nutrients* that your body needs. There are many different kinds of nutrients. Nutrients that give you energy are called carbohydrates. Nutrients that help to build your muscles are called proteins. Nutrients that protect your body organs are called fats. Fats are also a good source of energy. Minerals are nutrients that help make up some of your body parts. One mineral is calcium. Calcium is needed to build bones. Another mineral is iron. Iron is needed to make blood. Vitamins are used in different ways. For instance, vitamin A helps your eyesight. Vitamin C helps fight germs in the body. The chart below lists some nutrients. The chart also gives food sources for these nutrients.

NUTRIENT	FOOD SOURCE
Carbohydrates	Potato, sugar, bread, cereals, pasta
Proteins	Fish, chicken, eggs, beans, peas
Fats	Oil, lard, margarine, butter, nuts
Vitamin A	Carrots, apricots
Vitamin C	Tomatoes, oranges, strawberries
Calcium	Milk, cheese, yogurt
Iron	Spinach, liver

Go on to the next page.

Nutrients for Your Body, p. 2

✎Complete this nutrient puzzle.

Across

2. These foods give you energy.

3. This vegetable is a good source of vitamin C.

6. These nutrients help your body fight diseases.

7. This food is a good source of protein.

8. This vegetable is a good source of iron.

Down

1. This nutrient helps build strong muscles.

4. This food is a good source of calcium.

5. This food is in the fats group.

7. Margarine is a good source of these nutrients.

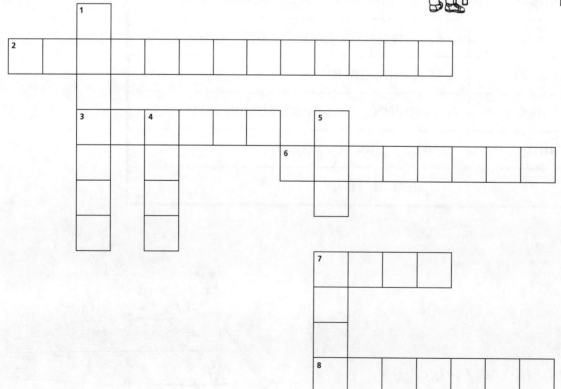

Unit 1: Nutrition
Health 3, SV 2705-7

Foods and Their Nutrients

· ·

   Each food has different nutrients that help the body live and grow. For example, milk has calcium. Calcium is a nutrient that makes bones and teeth grow strong.

✎**Look at the foods below. Write the name of the food under its nutrient.**

1. Carbohydrates   **2.** Fats   **3.** Vitamin C

4. Protein   **5.** Vitamin A   **6.** Calcium

Name _____ Date _____

Nutrition Scavenger Hunt

Food labels give information about the nutrients in foods. Look in the grocery store or at foods in the kitchen. What nutrients do foods have?

✎**Find as many foods as you can for each nutrient. Write the food name under the correct category.**

1. Carbohydrates

2. Fats

3. Vitamin C

4. Protein

5. Vitamin A

6. Calcium

Name _____ Date _____

Reading Food Package Labels

. . . .

It is important to know what is in the foods we buy. Learn what reading food package labels can tell you.

Materials:

three different cereal boxes

Do this:

• Find the Nutrition Facts chart on each box.

• Write the name of each cereal and the serving size in the chart.

Cereal Name	Serving Size
a.	
b.	
c.	

✎Answer these questions.

1. Which cereal has the most calories in one serving? _____

2. Which cereal has the most protein in one serving? _____

3. Which cereal has the most carbohydrates in one serving? _____

4. Which cereal has the most fat in one serving? _____

5. Which nutrients are listed on each box?

   a.   b.   c.

Go on to the next page.

Reading Food Package Labels, p. 2

6. Are there any vitamins listed that provide less than five percent of the

RDA (Recommended Daily Allowance)?_____

If so, which ones are they? _____

_____

7. Find the list of ingredients. Does any cereal list sugar as one of the

first three ingredients?_____

What does that tell you? _____

8. Which of these cereals has the most fiber in one serving? _____

9. Do you think any of these cereals are empty-calorie foods? _____

Why or why not?_____

_____

10. Which of these cereals is best for you? _____

Why? _____

_____

Fat Test

Do you know what foods contain fats? Fats are found in a lot of different foods. Sometimes you can't see fats. For example, there is fat in milk, meat, and eggs. You can find out which foods have fats.

Do This:

1. Get a brown paper bag, a pair of scissors, and some different foods to test.
2. Cut the paper bag into 3-inch squares. Write the name of each food you wish to test on a different square.
3. Rub a piece of food on a square until it leaves a wet spot. If the food is liquid, put a drop of it on the square.
4. Set aside the squares to dry.
5. When the squares are dry, hold them up to the light. If there is a greasy spot, the food contains fat. Fill in the chart below.

FOOD	FAT	NO FAT
Milk		

✎Answer these questions.

1. Were you surprised at which foods had fat? Explain.

    _____

    _____

2. Why do you think you should not eat foods with large amounts of fat?

    _____

    _____

Snacking Test

✎Take this test about snacking.
Check *Yes* or *No* next to each statement.

	YES	NO
1. I do not snack if it is close to mealtime. This can ruin my appetite.		
2. I brush my teeth or rinse my mouth after snacking. This helps prevent cavities.		
3. I do not snack on sticky, sweet food. These can stick to my teeth and cause cavities.		
4. I snack on healthful foods such as raw vegetables, fresh fruit, nuts, cheese, or even pizza.		
5. I do not snack on junk foods such as soda pop, potato chips, pretzels, cookies, and candy.		
6. I do not eat a snack if I am not hungry.		

Count up the number of yes's.
If you score: 5 to 6, you're a super snacker!
3 to 4, you're on the right track.
1 to 2, you need to make a
few changes.

Healthy Teeth

Your permanent teeth began to appear when you were about six years old. You will have most of them by the time you are 12 or 13. You want your teeth to be healthy and strong. You don't want to have cavities.

Cavities are caused by bacteria that are always in your mouth. These bacteria feed on the food that is stuck in your teeth. They form an acid that eats holes in your teeth. You can help prevent cavities. Here are some rules for teeth care.

1. Brush your teeth every morning and evening.

2. Use a toothbrush that has soft bristles and a flat top.

3. Visit the dentist twice a year.

4. Do not eat a lot of sweets. This includes chewing gum and soda.

5. Eat fruits and dairy products every day.

6. After eating, brush your teeth or at least rinse your mouth.

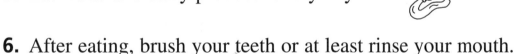
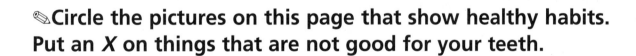

✎**Circle the pictures on this page that show healthy habits. Put an *X* on things that are not good for your teeth.**

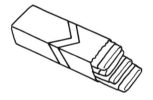

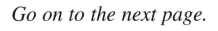

Go on to the next page.

Healthy Teeth, p. 2

✎Do you follow all these habits? Keep this chart for one week. At the end of each day, check off the habits you followed that day.

	Mon.	Tues.	Wed.	Thurs.	Fri.	Sat.	Sun.
Brushed my teeth in the morning							
Brushed my teeth in the evening							
Ate dairy products							
Ate fruit							
Brushed or rinsed mouth after eating							
Did not eat a lot of sweets							

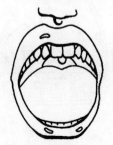

Calories

Your body turns the food you eat into energy. A *calorie* is the unit of measurement that tells how much energy food can make. Most food labels tell how many calories a food has. Many health and nutrition books list the number of calories in food.

List your five favorite foods on the chart below. Look in a book to find out how many calories are in one serving.

Food	Number of Calories (per serving)

Are you eating more calories than you need? Explain your findings.

_____

_____

_____

17

Name _____ Date _____

Energy

•••

✎**Write the word in the blank to make each sentence true.**

1. We get _____ from food.
    a. energy   b. work   c. sick

2. Your body uses food as _____.
    a. play   b. fuel   c. rest

3. We use more energy when we _____.
    a. read   b. sleep   c. run

4. Foods with more _____ give us more energy.
    a. fat   b. nutrients   c. calories

✎**Look at the pictures. Write *Yes* under the things that take lots of energy. Write *No* under the things that do not take much energy.**

5.

_____

6.

_____

7.

_____

Unit 1: Nutrition
Health 3, SV 2705-7

Trail Mix

Without doing any cooking, you can make a tasty snack that is good for you. This snack is made with seeds and fruits. It's called trail mix because people sometimes take it along when they go hiking.

Materials:

bowl

spoon

variety of nuts, seeds, and dried fruit:

shelled peanuts, walnuts or pecans,

shredded coconut, dried apricots

or apples, dates, raisins, dried currants,

sunflower seeds, toasted oats

milk (optional)

Do this:

1. Have an adult cut up the dried fruits.
2. Put about one half cup each of nuts, seeds, and dried fruits into the bowl.
3. Stir the mixture.
4. Take a handful and enjoy eating it!
5. You may also like to add some toasted oats to the trail mix. Try eating it this way with milk for a tasty breakfast.

✎Answer the question.

6. Why do you think people who are active need to eat healthy foods?

_____

_____

_____

Sunburns

Have you ever had a sunburn? Some people burn more easily than others, but no one is immune. A sunburn can be very painful. It is wise to avoid sunburns. They can damage the skin.

If you know you will be exposed to the Sun for many hours, be prepared! Remember that you can get a burn even on hazy or cool days. Keep yourself covered. Wear a hat. Apply a sunscreen before going out. Reapply every two hours or after swimming.

Cold compresses and aspirin can help relieve the pain of a severe sunburn. Calamine lotion also helps.

✎**Complete each sentence. Unscramble the letters to find the words.**

1. Apply a _____ before going out in the Sun.
   UNESCRESN

2. Wear a _____ in the Sun.
   AHT

3. You can get a sunburn on cool or _____ days.
   ZAHY

4. Sunburns can _____ the skin.
   AMDGAE

5. Cold compresses and _____ can help relieve the pain of sunburn.   **PSIAINR**

Name _____ Date _____

Insect Pests

••

Do you know these insects? They are common pests. There are things you can do to outwit them when you are outdoors.

Mosquitoes: Keep cool and bathe often. Wear light colors. Wear long sleeves and pants in the evening and early morning. An ice cube or calamine lotion can relieve itching.

Horseflies: Watch out on warm, sunny days! Wear light colors. Wash fly bites well.

Black flies: Black flies like wooded areas near water. They are active during the day. Wear light colors.

Honeybees and Wasps: Wear light colors. Avoid flowery prints. Do not wear hair spray or perfume. Do not walk barefoot. Do not wave your arms at bees and wasps. If stung, apply ice cube and baking soda paste. If you have severe local swelling or any other allergic reaction, get to the doctor at once.

✎**Get a book on insects. Find the names of these insects. Label them.**

1. _____ 2. _____ 3. _____ 4. _____

Unit 2: Safe Outside!
Health 3, SV 2705-7

Insect Pest Safety

1. Keep cool and bathe often.
2. Do not wear perfume.
3. Do not wear bright colors or flowers.
4. Wear long sleeves and pants in the evenings.
5. Do not walk barefoot outside.
6. Use an insect repellent.
7. Do not swat at bees. Move away slowly.
8. Do not leave picnic food out.

Look at the picture. Circle what is wrong. Then write the number of the rule being broken.

© Steck-Vaughn Company

22

Unit 2: Safe Outside!
Health 3, SV 2705-7

Mosquitoes

You might think that mosquitoes read travel advertisements. They seem to appear at all the lakes and ponds as soon as the weather is warm. Mosquitoes do not visit these places for pleasure. They are born there.

Mosquitoes lay their eggs in fresh water. The water can be in a quiet corner of a pond, in a ditch, or in an old tire. A puddle will do, too. When the eggs hatch, the young don't look like mosquitoes at all. In this stage of their life cycle, the young are called *wrigglers,* and they look like tiny caterpillars. Wrigglers can't fly like adult mosquitoes. They breathe through a tiny tube like a snorkel that sticks out of the water. Wrigglers find many tiny organisms to feed on in the water.

✎**Answer these questions.**

1. Why are so many mosquitoes found near ponds and lakes?

_____

2. What does a mosquito look like when it first comes out of the egg?

_____

3. What do young mosquitoes eat?

_____

After a time, wrigglers go into a resting stage, and then they become adults. All this happens in a few days. As adults, mosquitoes can fly. The females get their nourishment by sipping the blood of people and animals. Have you ever been bitten by a mosquito?

When mosquitoes bite you, you may get sick. Mosquitoes do not cause disease. They carry—from one person to another—the organisms that cause the disease. The drawing on page 24 shows the sharp mouth parts that allow the mosquito to cut into the skin. Look at the chart on page 24, which shows some diseases that mosquitoes carry from one person to another.

Go on to the next page.

Mosquitoes, p. 2

....

Diseases Spread By Mosquitoes
Yellow Fever
Malaria
Encephalitis

4. How are diseases carried from person to person by mosquitoes?

_____

_____

_____

5. What feature does a mosquito have that allows it to pierce the skin?

_____

_____

People try to control mosquitoes by spraying chemicals to kill them. These chemicals may harm birds and other animals. A safer way to control mosquitoes is to drain ponds and puddles where they lay their eggs. However, this may be harmful to other wildlife that lives in ponds. Other ways to control the pests include encouraging natural enemies, like certain flies, bats, or fish, or spreading diseases that kill mosquitoes.

6. Why is the practice of draining ponds to kill mosquitoes not always a good idea?

_____

7. What methods of controlling mosquitoes might be better for the environment?

_____

Healthful Hiking

Exploring outside can be even more interesting than learning about nature from books. You can hear the calls of the birds and see the colors of leaves. While hiking in the woods, you will also get fresh air and exercise. Hiking in the outdoors can improve both your mind and your body.

Although there's much to be gained by going outdoors, there are also risks. There are a few things to do before you go on any hike. The most important is to let others know where you'll be going. That way, if you should run into trouble, others will know where to look for you. Second, follow a trail. Even though you may think you'll be able to find your way back, the safest path is the one that's marked. Away from the path, you could wander for hours before finding your way back. If possible, take along a trail map. Third, always wear the right kind of shoes for the area of your hike. Most often, the best shoes are sturdy hiking boots with nonslip soles. Regular shoes do not support your feet, and they make it difficult to climb up rocky areas. Fourth, take along a container of water.

The next part of your planning should be thinking about the season and the possible changes in weather. For example, in the winter a sudden change in the weather might bring a cold rain or a snowstorm. Wearing many layers of warm clothes and packing a waterproof poncho can help you stay warm and dry. Your body also needs extra energy when you get cold. Taking along snacks, such as nuts and raisins, can give you the energy you need. In addition, the Sun sets early in the winter. Planning a hike that is short enough to be over before nightfall is very important.

Go on to the next page.

Healthful Hiking, p. 2

For a summer hike, you will need to wear sturdy hiking boots and layers of lightweight clothing. You can remove layers as the day warms up. In some places, you may want to wear long pants and shirts with long sleeves as protection from certain plants and insects. In some areas of the country, ticks and insects carry diseases. Proper clothing and insect repellent can help protect you from these pests. After your hike, you should check your body for ticks. If you find them, ask an adult or your doctor about removing them. Clothing can also help protect you from poisonous plants—such as poison ivy, poison oak, and poison sumac.

✎**Answer these questions.**

1. What four things should you do to prepare for any hike?

_____

_____

2. What are some risks in winter hiking? How should you prepare for each?

_____

_____

3. What are some problems that you need to consider when you plan a summer hike?

_____

_____

4. How would you plan a hike for this weekend? What would you need to wear? What would you take along?

_____

_____

Poison Plants

Many people get a rash from plants found outdoors. The most common plant to cause a rash is poison ivy. The sap in poison ivy causes a rash hours or days after touching it. You can also get a rash from clothes or pets that have touched the plant. Even smoke from burning poison ivy can cause a rash.

There are some things you can do if you touch poison ivy. Wash your skin and clothes with soap and water at once.

If you get poison ivy, it will last one or two weeks. It goes away even if not treated. Scratching the rash does not spread it, as many people think. You can buy calamine lotion at the drugstore to help stop the itching.

What does poison ivy look like? It is a woody vine or small bush. The leaves are bright green and grow three to a stem. In fall, it turns orange or red. The plant has small, greenish-white flowers. Later, it gets small, white berries.

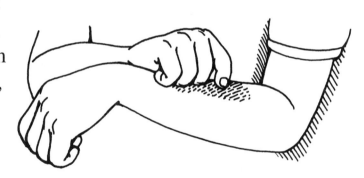

There are two other plants that cause a rash. They are poison oak and poison sumac. Poison oak also has three leaves on a stem, but its leaves are rounded. The leaves have short, soft hairs on the bottom. Poison oak has small, yellow flowers and small, white berries. It grows as a small shrub about three-feet tall.

Poison sumac has shiny, green leaves that grow 7 to 13 on a stem. It can be a small tree or a large bush. The stems are often reddish. Poison sumac has small, greenish flowers and gray berries.

Go on to the next page.

Poison Plants, p. 2

· ·

✎Use the descriptions from page 27 to identify each poisonous plant. Write the name of each plant under its picture.

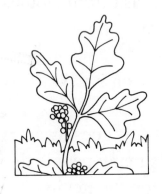      

1. _____ 2. _____ 3. _____

✎Answer these questions.

4. What can you do to prevent a rash when you hike?

_____

_____

_____

5. If you come into contact with a poisonous plant, what can you do?

_____

_____

_____

Plants–Harmful and Helpful

You may have learned the hard way about some harmful plants. If you've ever come back from an outing in the woods with an itchy rash, you may have learned firsthand about poison ivy, poison oak, or poison sumac. These plants have oils in their leaves that cause an allergic reaction in most people. If you brush against the leaves, your body reacts to the oil. First, you get a rash; then, it starts to itch.

Most poisonous plants are not harmful unless they are eaten. What may surprise you is that people eat poisonous plants by accident all the time. Think about small children. They will put almost anything in their mouths, and they often eat house plants. Chewing a single leaf of the popular house plant called *Diffenbachia,* or dumb cane, can cause a child to be unable to speak for a while. If larger amounts are eaten, the leaves can cause internal bleeding and damage to the liver and kidneys.

There are many other common poisonous plants. For example, hemlock, a flowering shrub with white flowers, was used as a poison by the ancient Greeks. If this plant is eaten, even in small amounts, it can paralyze or kill a person or an animal.

Many plants affect the heart. They either speed it up or slow it down. Often these effects are helpful. However, if the plant is eaten, the effects can kill a person. For example, foxglove is a beautiful flower that is often found in gardens. It contains something that is very helpful in treating heart conditions. However, if a person or an animal eats the plant, it can be deadly.

Go on to the next page.

Plants–Harmful and Helpful, p. 2

There is an evergreen tree that grows in Washington and Oregon called the *Pacific yew*. It is so poisonous that eating any of its leaves or seeds can cause the heart to stop. There are no symptoms for a doctor to treat—the person or animal that eats it dies quickly. But the bark of this tree has been found to contain a substance that can be used to treat some kinds of cancer. In fact, it is better treatment for these cancers than anything else that we know about!

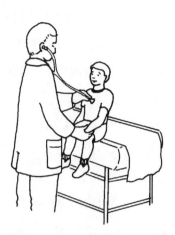

✎**Answer these questions.**

1. How can poison ivy affect people?

   _____

   _____

2. What are three poisonous plants? How do they affect you? Give examples.

   _____

   _____

3. What do the Pacific yew tree and foxglove have in common?

   _____

   _____

4. Why might it be a good idea to learn about the different kinds of plants that can be found in the woods before you go on a hike?

   _____

   _____

Thunderstorm Safety

Lightning is a kind of electricity. It can move easily in metal and water. It also moves through the body quickly, too. Lightning strikes tall objects. Here are some rules to stay safe in a thunderstorm.

1. Stay indoors.
2. Stay away from open windows.
3. Do not talk on the phone.
4. If you cannot get inside, do not stand under a tall tree. Try to get into thick woods.
5. Stay away from metal objects.
6. Get out of and away from water.
7. If you are in an open area, get down on your knees. Put your hands on your knees. Tuck your head.

✎**Draw a picture that shows how to stay safe in a thunderstorm.**

Water Safety Rules

Swimming is a fun way to play on a hot summer day. Here are some rules to help you stay safe around water.

✎**Unscramble the words in these rules.**

1. ___ ___ ___ ___ ___ how to swim.
E   R   L   N   A

2. Always ___ ___ ___ ___ with someone.
M   I   W   S

3. Swim only where a lifeguard is ___ ___ ___ ___ ___ ___ ___ ___.
T   C   I   A   W   G   H   N

4. Never call for ___ ___ ___ ___ unless you really need it.
E   P   H   L

5. Swim only in water that is ___ ___ ___ ___ for swimming.
A   F   S   E

6. When boating, always wear a ___ ___ ___ ___ jacket.
F   I   E   L

7. Do not ___ ___ ___ ___ into water if you do not know the depth.
I   D   V   E

Bicycle Safety Rules

Bike riding is fun. Here are some rules to help you stay safe when you ride a bike.

1. Always wear a helmet.
2. Ride on the right-hand side of the street.
3. Obey all traffic signs.
4. Use hand signals.
5. Have a light on your bicycle if you ride at night.
Wear light-colored clothing.

✎**Look at each picture. Tell how the person is staying safe. Use complete sentences.**

1. _____
_____
_____
_____
_____

2. _____
_____
_____
_____
_____

Bike Reflectors

∙∙∙

Materials:

red and black construction paper
scissors
silver glitter
glue
flashlight
bicycle reflectors
newspaper

Do this:

• Draw two shapes on red paper. Cut out the shapes. Place them on
  the newspaper.
• Put small dots of glue on one shape.
• Sprinkle silver glitter on the glue.
• Place both shapes on a piece of black paper.
• Turn off the lights. Shine a flashlight on the shapes.

✎Answer these questions.

1. Which shape is easier to see?

_____

_____

2. Which shape is like a bicycle reflector?

_____

_____

3. How can bicycle reflectors help keep you safe at night?

_____

_____

First Aid

First aid is the immediate care that is given to an injured person. Sometimes, if the injury is a small one, the right first aid may be the only care that is needed. Would you know what to do for a burn, a bruise, a small cut, or a deep wound?

Burns damage or destroy some skin cells. If the burn is small and does not have blisters, you can run cold water over it. You might also hold ice against the burned area. If the burn has blisters or is large, a doctor should be called. Even a small burn is bad if it has blisters.

Ice or cold cloths are good first aid for bruises. The cold helps reduce swelling and cuts down on the pain.

First aid for cuts is very important. A small cut should be washed with soap and water. You should always put a clean bandage over the cut to keep harmful germs out. If the cut is deep, you may have lots of bleeding. You will have to hold a clean pad tightly over the wound to stop the bleeding until you can get to a doctor.

Go on to the next page.

First Aid, p. 2

• •

✎ Complete this puzzle about first aid.

Across

2. _____ _____ is the immediate care given to an injured person.
5. For a small burn with no blisters, run cold _____ on it at once.
6. Put ice or cold cloths on a _____ to cut down on the swelling.
9. A clean pad should be held _____ over a deep wound until you get a doctor.
10. Ice on a burn or bruise will help cut down on _____.
11. If a burn is blistered or deep, the person should be cared for by a _____.

Down

1. Always put a clean _____ over a cut.
3. Sometimes, if the _____ is a small one, first aid is all that is needed.
4. You should put _____ or cold water on a burn.
6. If a small burn has a _____ , it is a bad burn.
7. A small cut should be washed with _____ and water.
8. _____ damage or destroy some skin cells.

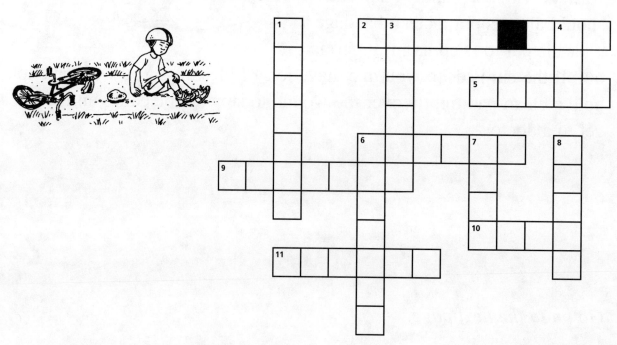

Name _____ Date _____

Exercise and Sleep

··

There are over 600 muscles in the body. All these muscles have different jobs. Some muscles help move blood through our bodies. Some move the food we eat. Other muscles help us breathe. Of course, muscles help us to hold things and to walk. If you do not use your muscles, they will become weak. That is why exercise is so important—it makes the muscles work and grow strong.

Exercise is important for another reason, too. Regular exercise can relax the body. It can help you sleep better. Sleep is an important part of keeping the body healthy. People need different amounts of sleep. Babies need the most sleep because their bodies grow so quickly. School children need about eight to ten hours of sleep. Adults need seven to eight hours of sleep. Sleep allows the brain and body to rest. If we don't get enough sleep, our brain and body do not work as well. Our attention wanders. Our muscles feel weak.

✎**Do you have healthy exercise and sleep habits? Fill in the chart below.**

Habit	Yes	No
I exercise every day.		
I stretch before I exercise.		
For short distances, I walk instead of riding in a car.		
I sleep eight to ten hours every night.		

Look at the chart. What can you do to become more healthy?

_____

_____

Warm Up and Cool Down

Have you ever been to a football game? Do you remember seeing the players warm up? Cold, tight muscles should be warmed up and stretched little by little. This prevents muscle injuries and muscle soreness. There are two types of warm-ups. General warm-ups are for the whole body. These exercises should take between three to five minutes. They should include some stretching, some calisthenics (like jumping jacks), and some walking and jogging. There are also specific warm-ups. These exercises help the body get ready for the sport. For example, in softball, players may throw the ball back and forth.

Cool-down exercises are just as important as warm-ups. Cool-down exercises take about ten minutes. You should gradually slow down your activity. For example, cool down after jogging by walking. Stretching should also be part of the cool-down exercises. This can prevent muscle soreness.

✎**Make a list of sports that you do. In the other columns, make a list of warm-up and cool-down exercises you can do.**

Sports	Warm Up	Cool Down

Exercise and the Heart

Exercise is important for everyone. If you do not use your muscles, you will become weak. Even astronauts must exercise. Since their bodies are not working against the full force of gravity, they hardly use their muscles in space. Bones and muscles get weak when there is not much gravity.

In 1970, three Soviet cosmonauts, who had been in space for 18 days, had to be carried from their spacecraft because their muscles were too weak to move their bones. Now astronauts exercise in space. They use a treadmill on which they run in place against a moving track. Using the treadmill makes their arm and leg muscles work harder. By using the treadmill for 15 to 30 minutes each day, they keep their bones and muscles healthy.

Working out on the treadmill is a form of aerobic exercise. Aerobic exercise forces the body to use a large amount of oxygen over a long period of time. If you are in good health, you should do some form of aerobic exercise at least three times a week for at least 15 minutes without stopping.

✎**Answer these questions.**

1. Why do astronauts need to exercise in space?

_____

_____

_____

Go on to the next page.

Exercise and the Heart, p. 2

2. How does the treadmill help astronauts?

_____

_____

Here's an exercise for you to try:

Place your hands against the wall at chest level. Try running in place while pushing for five minutes.

3. How is this exercise like running on a treadmill?

_____

_____

_____

✎**Here is an exercise chart for some students. The chart shows how the students' heartbeats changed after exercising for five minutes. Use the chart to answer the questions.**

Exercise Chart		
Student	**Heartbeats Before**	**Heartbeats After**
Tom	80	134
Lynn	75	130
Carlos	68	124

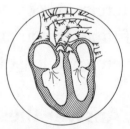

4. Who had the lowest number of heartbeats before exercising?

_____

5. Who had the lowest number of heartbeats after exercising?

_____

6. Whose heartbeats changed the most because of exercise?

_____

Good Hygiene

An important part of staying healthy is keeping your body clean. It is important to keep the germs and bacteria that can come in contact with your body from spreading. This is why you wash your body and hair. It is very important that you wash your hands whenever you handle garbage or raw meats (such as when you make a hamburger) and each time you use the bathroom.

Germs can spread easily from person to person, too. If you cough or sneeze without covering your mouth, germs fly out into the air and onto other people. If you cover your mouth, the germs are contained, and it is less likely that you will spread your germs to someone else. It is not a good idea to share straws, cups, combs, or hats, either. Germs can be passed from one person to the next this way, as well.

✎**Answer the following questions to see if you need to improve your hygiene.**

		Yes	No
1.	Do you take a bath or shower every day?	_____	_____
2.	Do you cover your mouth when you sneeze or cough?	_____	_____
3.	Do you wash your hands after using the bathroom?	_____	_____
4.	Do you wash your hands after handling garbage or raw meats?	_____	_____
5.	Do you share combs, hats, or other items that go in the hair?	_____	_____
6.	Do you share cups, straws, or other eating utensils?	_____	_____

If you answered *yes* to 1–4 and *no* to 5 and 6, good for you! You have good hygiene habits already! If not, try to improve your habits, and take the test again in two weeks!

Go on to the next page.

Name _____ Date _____

Good Hygiene, p. 2

•••

✎Look at each picture. Tell how it shows good hygiene. Use complete sentences.

1.

_____

_____

_____

_____

_____

2.

_____

_____

_____

_____

_____

3.

_____

_____

_____

_____

_____

Dental Hygiene

Eating the right foods is only part of what you can do to take care of yourself. You must be sure to brush and floss your teeth and see a dentist for regular checkups. These steps will keep your teeth from decaying. Decay is caused by acids in the mouth that eat the outer part of your teeth. The acids are caused by bacteria that live on the food in your mouth. If you brush and floss regularly, the food is taken out of your mouth, and the bacteria cannot live there. Decay can cause a hole in the tooth called a cavity. Tooth decay can also harm the gums. Regular dental exams and x-rays will detect any decay that you may have missed.

✎**Do you have healthy teeth habits? Fill in the chart below.**

Habit	Yes	No
Do you eat healthy meals?		
Do you eat healthy snacks?		
Do you brush your teeth after every meal?		
Do you brush your teeth at least twice a day?		
Do you floss your teeth every day?		
Do you visit the dentist regularly?		

✎**Look at the chart. What can you do to be more healthy?**

_____

_____

_____

_____

A Checkup

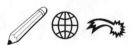

✎ **Look at the pictures. Write a paragraph to tell why it is a good idea to have a checkup once a year.**

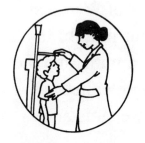

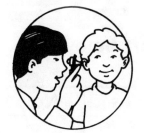

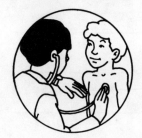

Name _____ Date _____

Germs That Cause Diseases

Diseases are spread from one person to another by germs that we cannot see. When germs enter your body, they multiply and cause disease.

What germs can make you ill? Bacteria are germs that are on everything you touch and in the water you drink. Many diseases, such as tuberculosis and blood poisoning, are caused by germs. Another kind of germ that can make you ill is called a virus. A virus is the smallest known disease-causing germ. Some viruses cause colds. A fungus is another germ that can make you ill. Fungi are usually found in damp places. Athlete's foot and ringworm are caused by fungi.

You can help protect yourself and others from diseases by getting vaccinations and by following good health practices.

✎**Locate the following words in the word search below. These words may be down, across, or on the diagonal.**

disease	germ	microbe	fungus
bacteria	virus	vaccine	health

F N V C D E K F D L E K J W J

V U A Z D Y J V I R U S W X A

P O C F U N G U S C O M I M L

M I C R O B E B E N G H R V U

S T I R G I H B A C T E R I A

A Y N M A B A O S P G Q R T S

X S E T L P Q H E A L T H M N

Using Bacteria to Stay Healthy

••

Have you ever had a vaccination? Did you ever drink polio vaccine? Doctors put bacteria into your body to keep you healthy. They use a weakened form of bacteria. The bacteria are too weak to grow in your body and make you sick. But the weakened bacteria make your body change. They cause your body to make *antibodies*. If the same kind of bacteria enter your body again, these antibodies protect you.

✎**Look at the chart. It tells what vaccinations many doctors suggest. Use the chart to answer the questions below.**

AGE	TYPE OF VACCINATION
2 months	diphtheria, whooping cough (pertussis), tetanus, [d.p.t.]; polio
4 months	d.p.t.; polio
6 months	d.p.t.
15 months	German measles, measles, mumps
18 months	d.p.t. booster; polio booster
4-6 years	d.p.t. booster; polio booster
14-16 years	d.p.t. booster

1. At what age are vaccinations first given?_____

2. What vaccinations are given at 6 months? _____

_____

3. What vaccinations are given at 15 months?_____

_____

4. Ask your parents what vaccinations you have had. Make a list of

them. At what age did you receive them? _____

_____

Name _____ Date _____

Changing Your Health Habits

✎Choose a health habit you would like to improve. Here are some examples. You can choose one of these or make up your own.

- I will floss and brush my teeth every day.
- I will not snack on sweets between meals.
- I will eat a good breakfast every morning.
- I will do vigorous exercise (such as running, swimming, or brisk walking) for 15 to 30 minutes at least three times per week.
- I will taste my food before salting.

1. What habit will you work on? _____

2. Think of a healthy reward you could give yourself after two weeks of doing what you said you would do. _____

3. Keep this record of your progress for two weeks. Write what you did toward improving your habit each day.

	Sun.	Mon.	Tues.	Wed.	Thurs.	Fri.	Sat.
Week 1							
Week 2							

Health
Grade Three
Answer Key

p. 3   1. C 2. A 3. B 4. B 5. C 6. A

p. 4   7. B 8. C 9. A 10. D 11. D 12. C

p. 5   Answers will vary.

p. 6   Answers will vary.

p. 8   ACROSS: 2. carbohydrates 3. tomato 6. vitamins 7. fish 8. spinach DOWN: 1. protein 4. milk 5. oil 7. fats

p. 9   1. Carbohydrates: cereal, bread, potato 2. Fats: nuts (also a protein) 3. Vitamin C: orange, strawberries
4. Protein: nuts, egg, turkey 5. Vitamin A: carrot, peach
8. Calcium: cheese, milk

p. 10   Check students' answers.

p. 11   Answers will vary.

p. 12   Answers will vary.

p. 13   1. Answers may vary. 2. Fats can be harmful to the teeth. Also, the body cannot use fats. The extra calories can make someone gain weight.

p. 14   Answers will vary.

p. 15   Circle: milk, girl brushing teeth, toothpaste, dentist, fruit, toothbrush; Cross out: soda, gum.

p. 16   Answers will vary.

p. 17   Answers will vary.

p. 18   1. a. energy 2. b. fuel 3. c. run 4. b. nutrients 5. Yes 6. No 7. Yes

p. 19   6. They need to eat healthy foods to give their body fuel.

p. 20   1. sunscreen 2. hat 3. hazy 4. damage 5. aspirin

p. 21   1. horsefly 2. honeybee 3. mosquito 4. wasp

p. 22   Students circle: girl with flowers (rule 3), boy with bare feet (rule 5), boy swatting bees (rule 7), food left out (rule 8).

p. 23   1. Mosquitoes hatch in water. 2. It looks like a tiny caterpillar. 3. Wrigglers eat tiny plants and animals found in the water.

p. 24   4. The mosquito picks up disease-causing organisms from the blood of an infected person or animal. It passes the organism to the next person or animal it bites. 5. The mouth parts are sharp and can cut the skin. 6. Other wildlife could be harmed. 7. It would be better to kill mosquitoes by increasing the number of natural enemies or spreading disease known to kill them.

p. 26   1. Tell others where you are going, follow a path, wear sturdy shoes and proper clothing, and take water. 2. For a sudden change in the weather, take along extra layers of clothing, a poncho, and a snack. Plan short hikes for days that get dark early. 3. Be aware of insects that carry diseases, and poisonous plants. 4. Answers will vary.

p. 28   1. poison oak 2. poison sumac 3. poison ivy 4. Wear clothing that covers the body. 5. Wash the parts of the body with warm water and soap.

p. 30   1. The poisonous oil in the plant causes an itchy rash. 2. Answers will vary. 3. Both are poisonous plants that are used as medicine. 4. It could help a person know which plants were safe or harmful to eat and which plants might cause a skin rash.

p. 31   Check students' drawings.

p. 32   1. learn 2. swim 3. watching 4. help 5. safe 6. life 7. dive

p. 33   1. The person is wearing a helmet and signaling that she will turn left. 2. The person is wearing a helmet and light-colored clothing while riding at night. The bike has lights.

p. 34   1. The shape with glitter is easier to see because it reflects light. 2. The shape with glitter is like a bike reflector.
3. Bicycle reflectors can keep you safe at night because they reflect the light from cars. Car drivers can see the bicycle rider better.

p. 36   ACROSS: 2 first aid 5. water 6. bruise 9. tightly 10. pain 11. doctor DOWN: 1. bandage 3. injury 4. ice 6. blister 7. soap 8. burns

p. 37   Answers will vary.

p. 38   Answers will vary.

p. 39   1. Without exercise in space, the astronauts' bodies do not work against the full force of gravity, and their bones and muscles get weak.

p. 40   2. It makes their arm and leg muscles work harder.
3. You do not move forward while pushing on the wall, but you exercise your leg and heart muscles against resistance.
4. Carlos 5. Carlos 6. Carlos's

p. 41   Answers will vary.

p. 42   1. The girl uses a tissue to sneeze or cough so that germs are not spread. 2. The girl is keeping her hair clean so germs will be removed. 3. The boy is washing his hands to remove germs before passing them on to someone else.

p. 43   Answers will vary.

p. 44   Answers will vary, but should include the need to make sure the child is growing and developing at a steady rate and getting vaccinations to prevent diseases.

p. 45

F	N	V	C	D	E	K	F	D	L	E	K	J	W	J
V	U	A	Z	D	Y	J	V	I	R	U	S	W	X	A
P	O	C	F	U	N	G	U	S	C	O	M	I	M	L
M	I	C	R	O	B	E	B	E	N	G	H	R	V	U
S	T	I	R	G	I	H	B	A	C	T	E	R	I	A
A	Y	N	M	A	B	A	O	S	P	G	Q	R	T	S
X	S	E	T	L	P	Q	H	E	A	L	T	H	M	N

p. 46   1. two months 2. d.p.t., or diphtheria, whooping cough, tetanus 3. German measles, measles, mumps
4. Answers will vary.

p. 47   Answers will vary.

Answer Key
Health 3, SV 2705-7